A New True Book

NIGHT BIRDS

By Alice K. Flanagan

Subject Consultant
David E. Willard, Ph.D.
Collection Manager of Birds at the Field Museum
of Natural History, Chicago, Illinois

CANCE

Children's Press®
A Division of Grolier Publishing
New York   London   Hong Kong   Sydney
Danbury, Connecticut

The earliest record of a barn owl dates back to France, about 19 million years ago.

In honor of the winged creatures that watch through the night

Library of Congress Cataloging-in-Publication Data

Flanagan, Alice.
   Night birds / by Alice K. Flanagan.
      p. cm. — (A New true book)
   Includes index.
   Summary: Explore the world of nocturnal birds and their activities.
   ISBN 0-516-01089-1
   1. Nocturnal birds—Juvenile literature. [1. Birds. 2. Nocturnal animals.] I. Title.
QL698.3.F57 1996   95-42386
598.251—dc20   CIP   AC

PHOTO CREDITS

Animals, Animals — © Stephen Dalton, cover, 8 (right), 14, 43; © Arthur Gloor, 10 (top left); © Gerard Lacz, 15; © John Gerlach, 17 (top left); © John Chellman, 17 (top right); © Zig Leszczynski, 17 (bottom left), 45; © Patti Murray, 17 (bottom right); © Bob Armstrong, 19 (right); © J. H. Robinson, 20 (top left); © Fritz Prenzel, 20 (bottom left); © Michael Dick, 23 (bottom); © Raymond A. Mendez, 24; © Souricat, 27; © Ray Richardson, 39

Jeff Foott Productions — © Jeff Foott, 23 (top left)

Valan Photos — © Ken Cole, 2; © Arthur Christiansen, 4, 8 (left), 20 (right); © Francis Lepine, 7 (left); © Wayne Shiels, 7 (right), 29; © J. R. Page, 10 (top right), 36; © Halle Flygare Photos, Ltd., 10 (bottom); © Anthony J. Bond, 12; © Bob Gurr, 19 (left); © K. Ghant, 23 (top right); © John Fowler, 26 (left); © Karl Weidmann, 26 (right); © Jeff Foott, 31 (left); © Robert C. Simpson, 31 (right); © Marguerite Servais, 32; © Wayne Lankinen, 35; © John Eastcott/Yva Momatiuk, 40

COVER: Common barn owl

Project Editor: Dana Rau
Electronic Composition: Biner Design
Photo Research: Flanagan Publishing Services

CONTENTS

WHEN NIGHT COMES

When night comes, everything in nature changes. The temperature drops, the sky grows dark, and a different group of animals begins to hunt and play. If we watch and listen carefully, we might hear and even see the birds of the night.

Usually, by nightfall, most daytime birds have settled down in the safety of their nests or in secluded roosts. Only a few birds might still be feeding in the light of the moon. Under the cover of darkness, night birds are waiting to replace the daytime hunters.

Nightfall brings busy activity. An owl silently swoops down and snatches a mouse. A male

Herons roost in flocks in trees or bushes, but they hunt alone along streams and shores.

nightingale serenades his "sweetheart." And a tube-nosed petrel dives for tiny sea animals that have risen to the surface of the water.

(Left) The song of the thrush nightingale is beautiful.
(Right) Barn owls are known to return to the nest about ten times a night with small animals to feed their offspring.

These are common nighttime activities. Yet, some owls also hunt during the day, and nightingales have been known to sing while the sun is up. Do you wonder why?

NIGHTTIME BIRDS AND DAYTIME BIRDS

Darkness brings safety from enemies, or predators. Petrels and shearwaters hunt at night when their natural enemies — hawks, gulls, and frigate birds — cannot see them. Flamingos, ducks, and swans remain active in the moonlight when people and predators will not bother them.

(Above left) Greater flamingos
fishing at sunset
A mallard (above right) and a
mute swan (below) feeding in
the moonlight

Birds that are active during the day are called diurnal. But with the aid of night-viewing equipment, sound recordings, and radio transmitters, we now know that many daytime birds are also active at night. A song thrush and a robin, for example, may sing well into the night if the moon is bright. A nightingale will sing by day, but it is best known for its evening song. And a gray heron may continue feeding

Two young
tawny owls wait
patiently in their
nest to be fed.

along coastal waters long
after the sun has set.

Some birds, however, are
active only at night. We call
them nocturnal. Most owls
are nocturnal. During the
day they remain well
hidden and asleep, but at
night they emerge to hunt
for their food.

OWLS, TRUE BIRDS OF THE NIGHT

In general, most birds are suited to a diurnal way of life. There are only a few groups of nocturnal birds. Owls make up the largest group.

Owls are perfect flying machines in the dark. They have keen night vision. Their large eyes allow them to see at a wide range, at great distances, and in low light.

Some owls can even turn their head sideways 270 degrees — almost all the way around. They also can turn their heads completely upside down — at 180 degrees.

An owl's sense of hearing is just as remarkable as its

The eagle owl is the largest of all owls. Its wings can span 6.6 feet (2 meters).

This tawny owl holds a deer mouse it has captured.

sense of sight. In total darkness a hunting owl can catch its food, or prey, using sound alone. The great gray owl's hearing is so sensitive that in winter it can hear the sound of a mouse under the snow.

Most owls have ear openings just behind their eyes. Feathers that stand straight up on some owls' heads may look like ears, but these structures have nothing to do with hearing at all. Scientists think that these feathers may help owls to distinguish each other at night.

Look at the outlines, or silhouettes, of the owls on the opposite page. If you saw their outlines in the dark, could you name each owl?

A screech owl (top left), Malaysian fish owl (top right), elf owl (bottom left), and long-eared owl (bottom right)

17

NIGHTJARS AND OILBIRDS

Nightjars and nighthawks are related to owls. They form the second major group of nocturnal birds. Special markings on the underside of their wings help us recognize them in the dark.

Nightjars hunt at dusk and on moonlit nights.

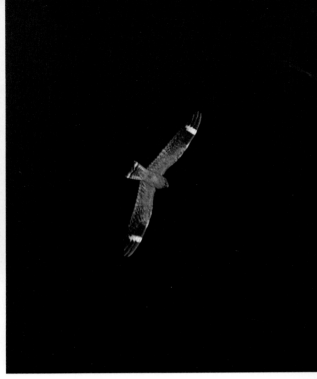

A nighthawk feeds on insects. When it is in flight, you can see its white bars on the underside of its wings.

Their huge mouths act like flying insect traps, collecting gnats and swarms of insects. Other birds that belong to this group are the tawny frogmouth and owlet

(Above left and
right) The nightjar,
or goatsucker, has a
jarring cry when it
feeds overhead at
night.
(Bottom left) Unlike
nighthawks and
nightjars, tawny
frogmouths feed on
ground-dwelling
insects.

frogmouth of Australia, and the South American oilbird.

Except for the oilbird, which lives in a cave in complete darkness, night birds must have some light to hunt and navigate. Skimmers and birds of the tube-nose order — albatross, petrels, and shearwaters — rely more on their sense of smell than on sight. As a result, they can find food and locate their nest burrows in the dark. The nocturnal kiwis of New Zealand also hunt by smell.

Oilbirds find their way in the dark by making a series of clicking sounds that echo off the sides of the cave walls and return to them. This is called echolocation. Swiftlets of Southeast Asia use echolocation, too. Swiftlets, however, are diurnal. They hunt for

(Above left) The black skimmer skims the surface of water for food.
(Above right) At the start of each breeding season, male shearwaters wait at underground nesting burrows for the females to arrive, usually at night.
(Bottom) The kiwi of New Zealand hunts for worms in the dark.

At night, some oilbirds may travel as far as fifty miles to find the ripe, oily fruit of palms and return before dawn.

insects by day and return to their caves at night, using echolocation to find their nests.

HUNTING BY NIGHT

Hunting is a major activity of night birds. Usually, their food sources are not available during daylight hours. Kiwis and woodcocks, for example, feed on earthworms that crawl to the surface of the soil only at night. Potoos and nightjars hunt flying insects that gather after

(Left) Woodcocks live in damp woods and bogs in many parts of the world. They eat earthworms and insect larvae. (Right) Potoos are related to nighthawks and nightjars. They live in trees and hunt flying insects.

dark. Owls hunt mice and small rodents that are night animals, too. Many shorebirds, especially the gray heron, fish at dusk and sometimes far into the night when the tide is low. Some small sea animals, called marine organisms, are more active at night and easier to spot. Tube-nosed petrels and shearwaters often feed on them.

Herons wade slowly in shallow water and hunt for prey. After catching it, they swallow it whole.

WHAT MAKES BIRDS ACTIVE AT NIGHT?

Why do some birds become active at night? They are responding to the changes in light and temperature. When the sun sets, the temperature drops and the air starts to hold more moisture. In other words, it becomes more humid. Lower temperatures and higher humidity bring

A mallard with its young

welcome relief to animals who are active at night, especially in the hot desert.

In many regions of the world, slugs, earthworms, and frogs emerge from the soil and their damp hiding places at night. Swarms of night-flying insects return to the air. As these tiny animals come out to eat, they become food for hungry predators.

A similar pattern of activity takes place on many

Oystercatchers (left) eat worms, crabs, snails, and shell-fish. The ovenbird (right) feeds on insects and worms. It is known for its beautiful night song.

waterways of the world at night. Along coastal waters and on the banks of ponds, rivers, and lakes, insects and tiny water animals become more active. Their appearance brings predators to the area.

Canada geese on a moonlit night in Ontario, Canada

BY THE LIGHT
OF THE MOON
AND THE STARS

The moon influences many nocturnal animals. It replaces the sun as a source of light. Night birds that rely on their sense of vision depend on the light of the moon to help them find food more easily. During a full moon, however, many night birds are actually less active. They wait for a new moon when the sky is darker and they are less likely to be seen by the animals they are hunting—or the animals hunting them. In some regions, birds that hunt insects also prefer to hunt during a new moon.

The gravitational pull of the moon controls ocean tides and makes coastal fish more available to birds. The activity of the

moon and the tides affects
the breeding activities of
animals that live near the
shore. It also increases the

A loon with its young

A Canada goose taking advantage of a well-lit night

number of birds trying to feed on them.

If the moon doesn't give off enough light, many nocturnal animals hunt by starlight alone. During bird migrations, or spring and fall trips to better climates, many birds are guided by the stars.

FOLLOWING THAT INNER CLOCK

Scientists believe that all living things have an inner clock that prepares them for day to night changes. They call it a "biological clock." No one knows for sure how this mysterious timing system works. It is generally believed, however, that it operates like an alarm clock. The clock keeps track of changes in the environment, such as day and night, the

Snow geese returning to a marsh as the sun goes down

Snow geese during their fall migration

seasons, the tides, and
the moon. Then it triggers
changes in plants and
animals to coincide with

changes in their environment.
For example, a biological
clock controls when an
animal awakens or sleeps.
It tells the animal when to
migrate, hunt, and mate.
It also prepares the animal
for the seasonal changes
that might threaten its
survival.

The biological clocks of
each species seem to be
timed to help that species
take full advantage of the
changes in its environment.

DAYBREAK

What happens as night begins to fade? As day breaks and the sun appears, nature once again returns to daylight activity. The temperature rises and daytime birds begin to sing. Barn owls go to roost — often in an old church steeple or the rafters of a barn.

A barn owl returning to its nest before daybreak

Oilbirds and swiftlets seek the darkness of their caves. And petrels and shearwaters fly back to their cliff-top burrows.

The winged night creatures of land and sea seek a place of rest and patiently wait for nightfall to come again.

A barred owl at rest in the swampy woodlands

GLOSSARY

bank (BANGK) — the rising ground at the edge of a pond, river, or lake

biological clock (by-uh-LAH-jik-uhl KLAHK) — an inner timing signal that keeps track of changes in the environment

breed (BREED) — to produce or increase by reproduction

burrow (BUR-oh) — a hole in the ground made by an animal for shelter or protection

coastal (KOH-stuhl) — on the land nearest the shore

distinguish (dih-STING-gwish) — to recognize by some mark or quality

diurnal (dy-UHRN-uhl) — active during the day

dusk (DUSK) — the period of time just after sunset, when little light remains

echolocation (eh-koh-loh-KAY-shun) — to locate something by making sounds that echo off an object

full moon (FOOL MOON) — the moon seen from the earth as a whole circle

gravitational pull (gra-vih-TAY-shuhn-uhl POOL) — a force of attraction that tends to draw particles or bodies together

humid (HYOO-mihd) — moist

marine organism (muh-REEN AWR-guh-nihz-uhm) — an animal that is found in the sea

mate (MAYT) — to join as partners; to breed

migration (my-GRAY-shun) — to move from one country or place to another at specific times

navigate (NAV-uh-gayt) — to steer or direct a correct course

new moon (NYOO MOON) — the moon's phase when its dark side is toward the earth

nocturnal (NAWK-tuhr-nuhl) — active at night

predator (PRED-at-ur) — an animal that kills and eats other animals

range (RAYNJ) — an area of land

rodent (ROH-dunt) — any of a group of mammals with sharp front teeth used in gnawing

roost (ROOST) — a support on which birds perch

secluded (sih-KLOOD-id) — hidden from sight

serenade (seh-ruh-NAYD) — to sing to

silhouette (sih-luh-WET) — an outline

tide (TYD) — the rising and falling of the surface of the ocean caused twice daily by the attraction of the sun and the moon

transmitter (trans-MIT-er) — a device that sends out radio signals

INDEX

(**Boldface** page numbers indicate illustrations.)

ABOUT THE AUTHOR

Alice K. Flanagan is a freelance writer and bird advocate. She considers her strong interest in birds, and a feeling of kinship with them, a symbol of her independence and freedom as a writer. She enjoys writing, especially for children. "The experience of writing," she says, "is like opening a door for a caged bird, knowing you are the bird flying gloriously away."

Ms. Flanagan lives with her husband in Chicago, Illinois, where they take great pleasure in watching their backyard birds.